AF812730

the crush math experiments

31-days of study strategies and simple habits to help high school and college students succeed in math, STEM, school and life

Allison Dillard

Happy Hypotenuse Publishing, LLC

Copyright © 2022 Happy Hypotenuse Publishing

All rights reserved. No part of this publication may be reproduced, distributed, or transmitted in any form or by any means, including photocopying, recording, or other electronic or mechanical methods, without the prior written permission of the publisher, except in the case of brief quotations embodied in critical reviews and certain other noncommercial uses permitted by copyright law.

Published by:

Happy Hypotenuse Publishing, LLC

Mission Viejo, CA 92691

Visit the author's website: www.allisonlovesmath.com

To contact the author, please email: allison@allisonlovesmath.com

The Crush Math Experiments: 31-days of study strategies and simple habits to help high school and college students succeed in math, STEM, school and life/ 1st ed.

ISBN: 978-1-950720-22-4

also by

Allison Dillard

The Math Anxiety Handbook

Abby Solves the Kitten Problem

Crush Hypothesis Testing

The Love Math Journal

Raise Your Math Grade

Crush Math Now

Available for Purchase on Amazon, Barnes & Noble, and other online retailers.

For a free copy of Raise Your Math Grade, visit allisonlovesmath.com.

contents

introduction

Dear Reader,

Why did I create *The Crush Math Experiments* for you?

Crush because I don't want you to timidly scrape by in math, barely pass, and end your class unconfident in your skills and abilities. I want you to *crush math*. I want you to walk away feeling powerful. I want you to have the confidence and success skills to learn anything and solve any problem. My goal is to help you figure out how to crush math, which will teach you how to learn complex things that you might think you cannot learn. It is an essential step in your quest to become the kind of person you will be proud of.

Math because that's the challenge you face today, the requirement your school or program requires of you. To become worthy of the degree or job you're striving for, you must become someone who can crush whatever is required of you, including math.

Just as sports and working out exercise and strengthen your muscles, learning math strengthens your mind, problem-solving skills, and ability to find patterns and make inferences. Yes, math is hard, but this book will help you learn how to learn math.

Experiments because the key to reaching any goal is taking action. Not just any action, though. Effective action. How do you know which actions are effective? By experimenting with different actions until you find the ones that get you the results that you want.

Each chapter provides an experiment for you to try out, so you can try new things, change up what you're doing, and learn how to learn math more quickly, effectively, and confidently.

My hope is that these experiments will help you beyond math, too. I hope that they will give you the power to learn anything difficult, anything intimidating, anything that you may need to learn on your quest to succeed in math, STEM, school or life.

Love, Allison

one
an experimental mindset

Before you get started, you need to know a little bit about mindset and how it affects your ability to learn new things.

Your **mindset** is your set of attitudes and beliefs.

Students with a **fixed mindset** believe their math skills are fixed. This discourages them from trying, practicing, and working hard in math. If you approach math with a fixed mindset, you might say or think things like:

- Some people just have a "math brain."
- I'm just not good at math,
- There's no point in trying. I'm just going to fail.
- Math is too hard.
- I'll never be good at math.
- I made a mistake. Therefore, I'm bad at math.
- I can't do it.

A more helpful way to approach math is with a growth mindset. Math students with a **growth mindset** believe their math skills can be developed through dedication and hard work. If you approach math with a growth mindset, you might say or think things like:

- My success does not depend on the skills I was born with.

- My success depends on the work I put in.
- Math is just a skill, and I'll improve with practice.
- All things are difficult before they are easy.
- Mistakes are a part of the learning process.

This is more helpful because it's true: Math *is* just a skill, and you *will* improve with practice. The trouble students with a growth mindset encounter is you can be hard-working and approach math with a growth mindset, but still get stuck. If you're doing the wrong thing, not getting feedback, and not learning from mistakes, you can get and stay stuck for a long time.

An even better way to approach math is with an **experimental mindset**. Math students with an experimental mindset have a growth mindset, but their focus is on experimentation. They believe their math skills will develop *faster* if they jump in, *quickly* try new things, and *experiment*. Experiment, reflect, learn, repeat.

Students with an experimental mindset believe that:

- Each new experiment will teach me something new and strengthen my knowledge or skills.
- Experimentation is the fastest way to gather data, feedback, or knowledge.
- The *more* I experiment and try new things, the faster I will improve.
- The *faster* I experiment and try new things, the faster I will improve.

Why are these mindset terms important?

Because whether you approach math with a fixed mindset, a growth mindset, or an experimental mindset has an enormous impact on your success.

The purpose of this book is to help you approach math with an experimental mindset, so that you will learn, improve, and succeed faster.

Crush Math Experiment: Pick your favorite growth mindset or experimental mindset thought above. (My favorite is *All things are difficult before they are easy.*) Write it down next to *nine* challenging math problems you'll need to study for your next test.

two
the crush math formula

The Crush Math Formula will help you to approach math with an experimental mindset. It breaks down into four steps:

1. **Set a Goal.** What is your goal for math? What are you working towards? Is it to raise your grade on your next test? To maintain your grade, but study more efficiently? To find more challenging problems to tackle? Whatever it is, get specific.
2. **Experiment.** What actions are you taking to reach your goal? What new strategies can you experiment with? Brainstorm as many as you can, then try *nine*.
3. **Reflect.** What experiments worked and what didn't? What's working, but needs to be tweaked or refined? What was effective? What was a waste of time? Review and learn.
4. **Repeat**. Continue this process of experimentation and reflection until you've met your goal.

That's it!

The beauty of the *Crush Math Formula* is that it's simple and easy to follow. The individual steps, however, require some honest self-analysis (but that's a valuable skill to develop) and some hard work (but working hard when it's effective is good).

Crush Math Experiment: Try the Crush Math Formula. Write down your goal for math and the actions you have been taking to achieve your goal. Circle the things that are working. Cross out the ones that are not.

three
the rule of 9

Comedians seem so gifted. Are they just born knowing what's hilarious? I asked a comedian how he knew what the funniest punchline to a joke was, and he said he didn't. He'd brainstorm nine different punchlines and pick the funniest.

Why nine?

Because it gets you past the generic and most obvious punchlines. It forces you to get creative and think outside the box; often, that's where the gold is.

Finding a punchline is similar to finding a solution to a math problem.

- There are many potential punchlines for a joke and there are many different ways to solve a math problem.
- One punchline might be funnier and one math solution might be more efficient.
- Comedians must experiment with different punchlines to find the best and math students must experiment with different solutions to find the best.

I challenge you to stop looking for one path to the correct answer in math. Instead, look for nine. By trying to find nine solutions, you'll understand the concept more deeply, strengthen your problem-solving muscles, and find the most efficient solution.

Crush Math Experiment: Pick a recent homework or test question that you can solve in more than one way. Write out nine different ways to solve it. Which is your favorite?

four
the 80/20 rule

The 80/20 Rule is not a math equation or even an official rule. It is more of an observation, but a helpful one when it comes to learning *how to learn* math.

The 80/20 Rule, also known as the Pareto Principle, was developed by an Italian economist, Vilfredo Pareto, who noticed that 80% of the land in Italy was owned by only 20% of the people. He also noted that 20% of the plants in his garden provided 80% of the fruit.

The 80/20 Rule says that if you try many different things, 80 percent of your results will come from 20 percent of the things you do. In math, this translates to:

- 20% of the people you ask for help might provide 80% of the clarification and explanations you seek.
- 20% of the questions you study might determine 80% of your exam grade.
- 20% of your mistakes might cause 80% of your wrong answers.

The most important 20% of people, questions, problems, mistakes, and study skills will be unique to you. If you experiment with many things, you will find the ones that will help you the most. Try out the experiments in this book and seek out the 20% that will provide you with 80% of the benefits.

Crush Math Experiment: On your last homework assignment, circle the 20% of homework problems that were the most difficult or took up the most time. Schedule extra time to study these for your next test.

five
imagine your future self

Imagine you love math. Imagine you have the skills, confidence, and grit necessary to succeed in any type of math and anything in life.

- You blaze through homework in record time.
- You rock your exams.
- You love the rush you get upon solving a complicated problem.
- You feel confident in your ability to learn new things, solve problems, and overcome challenges.

The sky is the limit now because math is in everything. Whatever your problem is, math is the answer. Whatever your goal is, math is a stepping stone to reaching it. If you look deeply enough, math is in everything. There is math in:

- Health and Fitness
- Sports and Economics
- Time Management and Prioritization
- Finances and Career Building
- Friendships and Team Building
- Marriage and Family
- In solving problems big and small to positively impact the world

For example, math gave me the skills to write this book. It helped me find time in my packed day to write. It helped me figure out how many words to write per day to finish by a specific deadline. It helped me develop the grit and discipline to write every day when there were so many easier things to do. It helped teach me that all things are difficult before they are easy, which is true for writing books and learning math.

If you approach every problem you ever encounter in life with a mathematical mindset, you will solve it. That is the beauty of math. Used correctly, it can elevate your whole life.

Imagine you love math and can solve any problem anyone throws at you. How would that change you, your life, and the future you envision for yourself?

Crush Math Experiment: Imagine you love math. Write down nine jobs you can imagine yourself doing or problems you can help the world solve.

six

how to get started on tough problems

One of the more stressful moments in math is staring at a test question and not knowing where to start. You know you need to write something down, but what? Everything you're thinking feels wrong. Or maybe you're not thinking about anything at all. Your mind is just blank.

What do you do? Use the Rule of 9 to brainstorm nine questions to ask yourself at the start of the problem. Then, work through the questions until you find one that moves you forward.

Here's an example. You're faced with a word problem that's confusing and convoluted. You're not even sure it's related to what you've been learning in class. Here are a few things you can do to start:

1. **Circle any numbers.** How can you use the numbers you are given?
2. **Circle variables and unknowns.** Have you solved something similar before?
3. **Highlight the question.** What homework or class questions asked something similar?
4. **Underline units of measurement.** Are there equations or rules associated with these units?
5. **Highlight keywords.** Can you use them to write an equation or find a relevant formula?

6. **Think back to recent homework or class problems.** Are there any similar questions that you covered?
7. **Think back to class.** Are there equations, theorems, or rules you've used or learned that could be relevant here?
8. **Draw a picture.** What can you include in a picture, and what is missing?
9. **Make a table or list.** What does organizing the values and variables into a table or list tell me?

The secret to being able to recall and work through these nine questions on an exam is to use them in homework and practice tests. Practice starting problems in different ways during homework for it to be easier to do on a test.

Crush Math Experiment: Find a hard homework problem that you haven't solved yet. Try starting it nine different ways.

seven

the fear of failure

I've made billions of dollars of failures at Amazon. Literally billions of dollars of failures... But they also don't matter. - Jeff Bezos, founder of Amazon

Some students are terrified of making mistakes, viewing it as proof that they're not good at math. Some students are so scared of making mistakes that they won't even try.

I challenge you to embrace Jeff Bezos's perspective. If he made 'billions of dollars of failures,' he must've tried many different things and failed many times. Likewise, in math, you should try many different things. If you make many mistakes, you have many opportunities to learn.

If Bezos thinks billions of dollars of failures don't matter, why do we believe that mistakes in math matter? If we think about it, what is the negative outcome that results from a mistake or getting a question wrong? That you get feedback and learn from it?

I agree with Bezos that these failures don't matter, and I hope you start to consider your mistakes and failures in a similar light.

Here are a few more epic failures worth noting:

- As a sophomore in high school, Michael Jordan didn't make it onto his Varsity basketball team.
- J.K. Rowling was once a jobless single mom fighting depression and living on unemployment.

- Abraham Lincoln lost elections for the state legislature, Congress (twice), the Senate (twice), and the Vice Presidency.
- Jack Canfield's Chicken Soup for the Soul book was rejected by 144 publishers.

One failed math test (or even 10 or 20) is nothing compared to these failures.

When you fail, pick yourself up, reflect on what you did, learn from your mistakes, and keep going. If you do this, you've not failed – you've learned something difficult and taken a crucial step in the journey toward success.

Crush Math Experiment: Write about a failure you faced in math or something else. Perhaps you failed a class or test. Maybe you got cut from your basketball team or had a math problem you couldn't solve. Pick a failure and write about nine things you learned from it.

eight
the pomodoro technique - part 1

Once upon a time, a college student named Francesco Cirillo couldn't bring himself to read his sociology book. He had an exam coming up and needed to focus, but he couldn't. There was just so much reading and work to do that studying seemed too big to tackle.

Then he had a realization.

Instead of focusing on the overwhelming task of reading the whole book and studying for his exam, he set his timer and read for ten minutes. When the timer went off, he realized he'd focused! He'd made progress! He moved forward and successfully prepared for his test by breaking his studying into bite-sized chunks.

Today, Cirillo's study habit is a worldwide phenomenon called The Pomodoro Technique. It is where you set your timer for 25 minutes and focus on completing one task only for 25 minutes.

I wrote much of this book in 25-minute sprints between 5 am and 7 am. Some teams at Google break their day into 25-minute sprints. And breaking your math homework up into 25-minute sprints can also make the task more doable and your time more effective.

Crush Math Experiment: Try the Pomodoro Technique nine times over the next week and see which assignments it helps.

nine
how to prepare for a swim meet

What is the best way to prepare for a swim meet? Is it to watch swimming videos on YouTube and review notes on how to swim?

No!

The best way to prepare for a swim meet is to jump in the water and swim. You'll never be sufficiently prepared for a swim meet if you never get in the water and swim.

Yet, many students use this strategy (watching videos and reviewing notes) to prepare for math tests. Watching, reading, and reviewing – these are passive forms of learning. They are ineffective. They can take up lots of time yet still leave you unprepared for your math test.

What's a better way to study for a math test? What's the equivalent of jumping into the water and swimming?

Test yourself.

Pick a few of the problems you've done in homework or class. Write each one on a blank paper. Pick random papers and solve them as you would on a test (i.e., without looking at the solution, your notes, your book, etc.). Then, correct it.

If you got them all correct, great! If you didn't get the questions correct, review them and note your mistakes. If you don't understand the solution, try to figure it out or ask for help. Move on to studying the next topic or more challenging questions.

The best way to prepare for a swim meet is to practice swimming as if in a swim meet.

The best way to prepare for a math test is to practice solving math questions as if you were taking your math test.

Crush Math Experiment: Pick nine questions from your homework or class notes. Write each one on a blank paper. Choose papers randomly and solve the questions. Review the ones you missed. Repeat as needed.

ten
average of five

You are the average of the five people you spend the most time with. - Jim Rohn, author of *Strategies for Wealth and Happiness*

If your five best friends are professional basketball players, there is a good chance you also play, are good at, or know a lot about basketball. If your five best friends are animal conservationists or professional piano players, odds are you know a lot about those things, too.

There are exceptions to every rule, but I like how Jim Rohn's average of five quote highlights the impact our closest friends have on us.

Are you surrounded by people who hate math? Or think math is too hard or too pointless? Or aren't people you can ask for help? If you are, it could benefit you to add some new people into your life – people who value math, work hard to succeed in math and can help you succeed in it, too. Build a team, a network, a support system.

Why is it essential to have a team?

- **Imagine you make a mistake.** If you're studying alone, do you even notice? If you don't, you keep making the same mistake over and over. If you're studying with others, someone might catch it, you learn from it immediately, and you move forward correctly.
- **Imagine you are stuck.** If you're studying alone, you might stay stuck and give up, or it takes forever and is frustrating. If

you're studying with others, you discuss it, bounce ideas off each other, and work through the problem-solving process together.

- **Imagine you don't like math.** If you're studying alone, you keep on not liking it. If you're doing homework with others (who enjoy math), their enthusiasm makes it more enjoyable.
- **Imagine you get a question right.** If you're studying alone, that's it. If you're studying with a group, you can explain it, which improves your understanding and communication skills.

When math gets hard, insisting on learning it alone is the long, slow route where you might get stuck. Studying with a team means figuring it out faster, getting feedback faster, and learning faster.

Now for the hard part. How do you find five positive, encouraging, math-crushing peers to study math with? Approach building a team with an experimental mindset. Learning how to build a team is a skill. Try reaching out to and working with lots of different people. If they say no, move on. If you work together and it's not productive or helpful, move on. Experiment fast. Learn fast. Build your team fast.

Crush Math Experiment: Think of nine people who might be able to help you in math. Send them a simple text or email saying, "Any interest in studying for the math test together?" or "Did you understand the homework question about...." Keep an eye out for the ones who are the most helpful.

If you don't need help in math right now, think of nine people you can help. Send them a simple text or email saying, "Any interest in studying for the math test together?"

eleven
focus on you

It's easy to fixate on the things we don't control or fall into blaming others. Things you don't control:

- How your teacher teaches
- How your teacher structured your class or grades
- When your test is
- What will be on your next quiz or test
- What information you are given to help you prepare for your quiz or test
- What you got on your last test
- Any other thing that happened in the past

External factors are out of your control. Things that happened in the past are now out of your control. You cannot change them, so it's a waste of time and energy to dwell on them.

How do you overcome external factors and past events? How do you succeed despite them? Focus on you. "You" includes the things that you control, so focus on the following:

- What you do to study math
- How you study math
- When you study math

- Who you study math with
- How you approach math and think about math
- What changes you can make

Please don't get offended when I recommend that you focus on changing you. Ultimately, the thing you control, the thing that you can change, is... you. There isn't some written document that delineates who you are. Who you are is ethereal. It's malleable. It is ever-changing and, from the day we're born, we are meant to change for the better every day of our lives.

Crush Math Experiment: Be honest with yourself. If you're struggling in math, pinpoint nine changes you need to make to crush math. If you're already crushing math, help a friend who is struggling to pinpoint nine changes they can make to how they learn math.

twelve

how to learn from mistakes

If your teacher tells you to learn from your mistakes, how do you actually do that? Let's start with what not to do:

- Do not ignore the mistake or problem you couldn't solve.
- Do not tell your teacher you understand your mistake when you do not.
- Do not do nothing and secretly hope it does not come up on the following quiz or test.

It's frustrating to make a mistake on homework, then make the same mistake on your quiz, then make the same mistake on your test, then find out that you also need it for your next homework, quiz, and test.

If you never learn from it, one mistake can come up again and again.

But how do you learn from mistakes? After all, if you could figure it out, you wouldn't have made the mistake in the first place.

Here's how:

- Figure out how to solve the problem.
- Write out the work correctly.

If you cannot do both of those things, you haven't learned from your mistake.

The hard part is figuring out how to solve the problem. How do you do this? Tackle it like any difficult question, but remember that you're not starting from scratch. You have a lot to work with to help you figure it out. Here are my Top 9 questions to ask yourself to learn from your mistakes more efficiently:

1. **Were you given feedback?** If your teacher or the computer program (or anything or anyone) gave you feedback, start there!
2. **Were you given the correct answer?** If you weren't given feedback but were given the right answer, start there. Use the answer to your advantage. Can you work backward from the correct answer to figure out the steps or your mistakes?
3. **What are your common mistakes?** If you've been tracking your errors (which you should), check if any are relevant to the problem.
4. **Was there any information you missed?** Read through the problem again. Sometimes, we miss a detail. If there's part of the question you didn't understand or accidentally skipped over, consider how that missing piece of information affects the problem.
5. **What concepts, formulas, or theorems are relevant?** Review the concept, formula, or theorem related to the question. Did you miss part of it, misunderstand it, or apply it to the wrong problem?
6. **Can you find a similar problem?** Look back through your homework, class notes, or textbook for similar questions. Study the setup for the question, the steps, and anything else you can learn from it.
7. **What details are different?** Sometimes, the correct answer is hidden in the details. If you found a similar problem, but it's not quite the same, ask yourself what's different and think about how that affects the problem.
8. **Is there more than one way to solve the problem?** Often there is more than one path to the correct answer. Try stepping away from your incorrect solution and solving the problem in an entirely different way.

9. **Is there anyone who can help you?** If you can't figure it out alone, don't ignore it. This is when you need to ask for help. Go to your teacher, show them your work through the previous questions above and ask for help. Write down what they say and show them what you wrote to confirm that you understood correctly.

Crush Math Experiment: Find a question you got incorrect or couldn't solve. Work through the nine questions to solve it.

thirteen
motivate yourself

Digital marketing experts estimate that most Americans are exposed to around 4,000 to 10,000 ads each day. - Forbes

We live in a 3-second world of, Do this! Buy that! Look at this! Life is too short, my friend. Filter it all out – all the advertisements, all the 2-second look at my perfect life posts, all the noise.

Instead of letting the world decide what messages you'll see all day, hand-select specific messages that will positively impact you. Keep those messages where you'll see them frequently. Here are some quotes that you might find helpful:

- *It's not that I'm so smart, it's just that I stay with problems longer.* — Albert Einstein
- *Be intrigued by mistakes, enjoy effort, and keep on learning.* — Carol Dweck
- *It does not matter how slowly you go so long as you do not stop.* — Confucius
- *Whether you think you can or you think you can't, you're right.* — Henry Ford
- *Our greatest weakness lies in giving up. The most certain way to succeed is always to try just one more time.* — Thomas Edison
- *You must do the thing you think you cannot do.* — Eleanor Roosevelt

Over the years, I've experimented with where to put the messages I need to see. For my favorite quote, "All things are difficult before they are easy," I find it helpful to put it on the wall by my desk. I probably see this 100+ times a day. I also set it as a daily reminder on my phone, so I see it once a day, no matter where I am.

It's strange how this message has impacted me in many ways:

- Every new level of math I took in school was difficult before it was easy.
- Writing books was difficult before it was easy.
- I gained some weight during covid, and getting back to a healthy weight was difficult before it was easy.
- This fall, I hosted my first conference – for 1,200 math teachers – and WOW, was it difficult, but I know the next one will be easier.

Seeing a message repeatedly can affect you positively in so many different ways.

Crush Math Experiment: Find a quote that motivates, inspires, or reminds you of something you need to remember. Put it in nine places, where you will see it throughout the day.

fourteen
reframe problems as questions

It's easy to get stuck in negative thoughts. They can easily make us feel overwhelmed or powerless. Instead, focus on asking questions that allow you to consider what changes to make.

Every problem, every fear, and every negative belief can be rephrased as a question. For example, the negative thought, "I can't solve this problem," can be reframed as:

- What are nine things I can do differently to solve this problem?
- What are nine things I am given to help me solve this problem?
- What are nine similar problems that can help me solve this problem?
- What are nine strategies I've learned in class related to this problem?
- Who are nine people I can ask to help me solve this problem?

Another problem some students have is generally thinking, "I'm just not good at math." I would rewrite this as, "What are nine things I can do to become better at math?" Or if your problem is that you have Dyscalculia, think, "What are nine things I can do to overcome Dyscalculia to succeed in math?"

Reframing problems as questions reminds us that problems are just questions that need solutions. Like in life, some of your math problems are hard to solve than others, but ultimately they all have solutions.

Don't stay stuck.

Get in a growth, problem-solving, and experimental mindset. Reframe problems, fears, and negative thoughts as questions just waiting for you to find the solution.

Crush Math Experiment: Write down a negative thought you've had about math or something else and turn it into a question (or nine questions).

fifteen

how to finish your homework

Eat the frog.

Yes, you read that correctly.

Mark Twain once said, "If it's your job to eat a frog, it's best to do it first thing in the morning. And if it's your job to eat two frogs, it's best to eat the biggest one first."

If math is your frog—the job you hate and must complete—then it's best to do it first.

- Do your math homework first, so it doesn't hang over your head, ruining the rest of your day.
- Do your math homework first, when you have the energy, patience, and willpower to get through a mentally challenging assignment.
- Do your math homework first, even before you do your other homework.
- Do your math homework first, so if it ends up being harder than expected, you have time to revisit challenging problems or ask someone for help.
- Do your math homework before watching TV or playing video games, so you can enjoy those things without thinking about the frog you still have to eat.

People who go to the gym first thing in the morning, no matter what, get their workouts in more consistently than those who go at the end of the day if they have time. Writers who write first thing in the morning, no matter what, publish more consistently than those who write in the evening after everything else is done. Do your math homework first to get it done more consistently.

Working earlier, not harder, is the key to getting math done. Do the hard work first to do the easier things and have guilt-free fun later. Eat the frog and crush math.

Crush Math Experiment: Brainstorm the earliest nine times this week that you can get math done.

sixteen
the pomodoro technique - revisited

When you practice soccer, you practice for a set time and then stop. You don't go until you're exhausted and frustrated and hating soccer. If you did that, everyone would hate soccer.

When you practice piano, you practice for a set time and then stop. You don't play until your hands cramp up and you want to punch the piano. If everyone did that, everyone would hate playing the piano.

Likewise, when you study math, it can be helpful to study for a set time and then stop. Don't study math until you're exhausted, frustrated, and hating math. Pomodoros (i.e., working in 25-minute sprints) will help you study math because they help you:

- **Estimate how long things take.** When you write down what you will work on, you estimate how much work you can do in 25 minutes. When your timer goes off, you see how much you got done.
- **Focus better**. If you know you're only studying for 25 minutes, you can power through, stay focused, and get more done in a shorter time.
- **Study faster**. Think of how you work during tests when the timer is running, and you know you must finish within a set time. You work faster and harder. This is because it's timed.
- **Create productive habits**. If you have 25 minutes of math and

you're done, this is a doable amount of studying. It's not too stressful to get started. It's easier to do it. And that's what you want when creating a habit: something that's doable.

The last step in the Pomodoro Technique is taking a 5-minute break, and it's equally important. After your Pomodoro is done, give yourself a 5-minute break. Get up and walk around. Relax. Give your mind a rest, and don't think about math. This offers several benefits, including

- **Helping you to manage distractions.** By giving yourself five minutes to check your phone notifications during your break, it'll be easier to ignore them during your Pomodoro.
- **Allowing ideas to percolate.** You might remember a formula or a similar problem from class. Often, the five-minute breaks are all you need for inspiration to hit.
- **Helping you manage stress.** Thinking about everything else you must do while simultaneously doing math makes math stressful. The Pomodoro Technique can help make math less stressful by cutting out distractions and studying faster.

Crush Math Experiment: Get out your math homework, set your timer for 25 minutes, and go! After the timer goes off, give yourself a 5-minute break. If you have more math to complete, repeat.

seventeen
find your professor towse

One of the reasons I majored in math, became a math professor, and write math books is because of my advisor at Scripps College: Professor Towse.

Why?

Because he taught me what all the math professors at Scripps College teach:

- To love math and the challenge of math
- How to learn math and difficult things
- How to pick yourself up when you fail or feel overwhelmed
- How to focus on the learning journey (and enjoy it) rather than just the result

I know that not every math department is as positive and encouraging as Scripps', but at least one Professor Towse is hidden within every math department. Even if most of your math department seems stuffy, distant, and discouraging, I challenge you to find a professor or teacher who:

- Will encourage you and help you.
- Believes you can succeed in math and wants to see you succeed.
- Will help you pick yourself up when you fall or find your way when you're lost.
- Help you learn how to learn hard things, like math.

As strange as it might be, analyzing and changing the educators you go to for help can impact your long-term success in math. Spend more time with educators who have a growth mindset. Make those your go-to people for questions and advice.

Crush Math Experiment: Make a list of nine educators you could go to for help in math. Reach out to each for help on a difficult math problem. Who is the most helpful and the most encouraging? Odds are, at least one will be, and now you've got an expert in the field to go to for help when you need it.

eighteen
how to study efficiently

What are the most important things to study for your next math test? To figure that out, ask yourself, "What questions or concepts am I afraid to see on the test?" and practice those.

What questions am I afraid of? This is the crucial question. Be honest with yourself. Don't just skim through your notes, thinking about and reviewing all topics equally. Make a list of specific questions from your homework, notes, quizzes, and textbook that you are afraid to see on your test. Your list might look like this:

- All of Friday's homework
- All of Section 3.4
- The word problems from Section 3.2

A list like this tells you that you might need to allocate extra time to study or ask for help to get through it more quickly.

Your list might also look something like this:

- Friday class notes - Example 5
- Wednesday's homework #26 & 27
- Section 3.1 - the last word problem

Even though it's a shorter list, how much time should you allocate to these challenging problems? Will you need to ask for help? Plan accordingly.

Crush Math Experiment: Make a list of nine questions you're afraid to see on your next test. Create a plan to figure out how to solve these problems and how to practice them for your test.

nineteen

how to overcome math anxiety

Math anxiety is when the tension and anxiety you feel towards math is so bad that it interferes with your ability to solve math problems. Some sources say that as much as 50% of Americans have some level of math anxiety!

If you have math anxiety, it's imperative to approach learning math and taking tests with a growth mindset. Just as learning math is a skill and you'll improve with practice, overcoming your math anxiety is a skill and you will improve with practice.

The key to overcoming your math anxiety is to tackle it head-on. I suggest starting with the following two suggestions from Dr. Maria Ryan, my co-author of the Math Anxiety Handbook:

#1 Reflect on your math anxiety origin story. Maria's research found that reflecting on what event of experience caused your math anxiety can be helpful. Remember that math anxiety isn't who you are. It's simply a feeling that started when you had a negative experience (or experiences) with math. Try reflecting on the experience that caused your math anxiety and see if it helps.

#2 Find a significant person. Overcoming math anxiety is difficult to do in isolation. Another part of Maria's work discovered that finding someone who can help you with math can have a huge impact on helping you to overcome math anxiety. Try finding someone who can help you succeed!

These two strategies are helpful for overcoming any type of performance anxiety. I had a fear of public speaking for years. Reflecting on where my public speaking anxiety came from and joining a weekly group called Toastmaster's helped me to overcome it.

Crush Math Experiment: If you have math anxiety, reflect on where it came from and brainstorm nine people who can help you to overcome it. If you don't have math anxiety, brainstorm nine people who might benefit from learning these two tips and share with them.

twenty
advertise to yourself

Let's take a look at two students, one who advertises to herself and one who does not.

Prisha gets out of bed and brushes her teeth. Next to the mirror is a note she wrote, "A+ on Calculus final." She goes to her closet to get dressed, and on the back of the door is a copy of a (pretend) doctoral certificate with her name on it. She had printed it up to remind herself of what she was working towards. She looks at her phone and sees the Dartmouth College emblem in the background. She eats breakfast and opens a pretty journal with "Stop doubting yourself, work hard, and make it happen" inscribed on the cover. Inside, she reviews her game plan for getting an A+ in Calculus and the steps she'll take today to work towards her goal. She finishes up and heads to school.

Saanvi also gets out of bed. While scrolling through Instagram and getting distracted down a rabbit hole of articles and social media posts, she brushes her teeth and gets dressed. Still glued to her phone, Saanvi eats breakfast and heads to school.

Prisha and Saanvi both have the same goal, to get an A+ in Calculus and go on to become a doctor. Based on their morning routines, who is more likely to achieve their goal?

Prisha.

Why?

Because she takes control of the things she reads and hears every day. She ensures that her dreams, game plan, and important messages are visible and frequently seen. She advertises her goals to herself, so she keeps track of them.

Saanvi, on the other hand, isn't paying attention. Saanvi's goals are out of sight and the messages that she hears are distractions from the rest of the world. She's more likely to forget or do fewer of those tiny things she needs to do daily to reach her goals.

Successful students understand that it's not the outside world's job to motivate or remind them of their goals. They know they must create this kind of environment for themselves.

Crush Math Experiment: Put your list of goals in nine places where you'll see them. Keep them there for a week to see which place(s) are the most helpful.

twenty-one
how to stop careless errors

So often, students and teachers chalk little math mistakes up to careless errors. As a student, it's frustrating because if it's careless, then there's nothing you can do besides be more careful next time, and that's super general advice.

But what if your careless errors are not due just to carelessness? What if an underlying cause makes you more likely to make certain mistakes? You'd want to know what that underlying cause is and fix it, right?

I'll be honest here: If you make careless errors, some of them might not be careless. Careless errors sometimes stem from gaps in your foundations that need to be fixed. What are gaps in foundations? Things you learned a long time ago that you still use in math today:

- Adding, subtracting, multiplying, dividing
- Positive and negative numbers
- Fractions, decimals, and percentages
- Following the order of operations

Right now, I know some students are thinking, "Baloney!! I wrote down -2 x 3 = 6 on my last test, but I *obviously* know that -2 x 3 = -6. It is so basic! It's insulting to suggest that I don't know such a basic fact. It was obviously a careless error."

Here's the thing: If you tend to make mistakes with multiplication or with positive and negative signs or the order of operations, then you are not alone. Many students understand their test corrections, but need to learn their basic math skills better to avoid careless errors.

What's the solution? Go back and practice the basic math skills causing you to make mistakes. Don't go back and read about those skills. Do practice problems where you multiply or add positive and negative numbers. Do practice problems following the order of operations. Do deliberate, repeated practice on those skills related to your mistakes.

Reviewing math skills from previous math classes can be humbling, but know that you are not alone. I've met many students who benefited greatly from brushing up on skills from previous math classes.

Crush Math Experiment: Write down the last nine mistakes you've made on homework questions or test questions. Are any of the mistakes similar or repeated? If so, do nine practice problems on those skills.

twenty-two
math is like soccer

I am not a "natural" athlete. My parents signed me up for soccer when I was a kid, and the truth is that, no, I was not a "natural." Some kids naturally chased the ball, were fast and aggressive, and walked onto the field the first day instinctively knowing how to dribble a soccer ball.

When I first walked onto the field, even running didn't come naturally. It took me longer than many other kids to learn basic skills like kicking, dribbling, passing, or sensing where the ball might go next on the field.

Even though I'd never become an Olympic soccer player, I eventually played on all-star teams and my high school JV team. I had health problems in high school, so I'll never know if I would have made Varsity. Still, I learned enough to enjoy playing in co-ed soccer leagues throughout my twenties and to coach my kids' soccer teams in my thirties. I learned enough to reap all the benefits of learning and enjoying a sport.

Why am I telling you this? Because math is like soccer. It is harder for some students than for others. And just because it is harder for you doesn't mean that you can't learn it or that you won't benefit from it. Even if you aren't aiming to be the best mathematician in the world, you will still benefit from the experience and skills learned from your math class. Like soccer, figuring out how to learn math when it's challenging for you is an important and worthwhile journey.

If you don't feel like math is one of your strengths, that's okay. Remember that learning math is the same as learning any skill like soccer, the piano, or a language. It will be harder for some than others, but like all skills, you'll improve with practice, and learning something difficult is worthwhile.

Crush Math Experiment: Brainstorm nine skills, besides math, that you have developed or are developing? Celebrate them! Not all of our successes or goals need to be centered around math. :)

twenty-three
an imbalanced life

It’s easy to lose track of your goals and get lost in a noisy world trying to pull your attention in different directions. It's easy to get sidetracked and to drift. It's easy to get caught up in urgent things rather than important ones. It's easier to take the easy route over the hard one. It's easy to meander and not grow up into the person you always assumed you'd become.

How do you keep yourself focused and moving toward your goals? By taking a deeper look at your goals and prioritizing.

Write down all your goals. Whether you have five or fifty goals, write them down. Academic goals, extracurricular goals, travel goals, relationship goals, financial goals, and work goals.

- It can be as big as “be a millionaire” and as small as “show up on time to class tomorrow.”
- It can be a repeated goal, like “check my phone less than fifty times a day” or “always do my math homework first.”
- Or it can be a one-time goal, like “go rock climbing” or “make the varsity team.”

Done? Good!

Why the list? Why write down all the goals? Because nothing comes without sacrifice. There are only so many hours each day, so when you choose to put time toward one goal, you must acknowledge that you're sacrificing time for another. There are always sacrifices to be made, so think about them and make conscious decisions about what you do and do not want to sacrifice.

Crush Math Experiment: Of all your goals, pick your top three. Write them down where you'll see them every day. Try to take one action every day that moves you toward each of your top three goals.

twenty-four

how to check your work

Teachers always say, 'Check your work,' but what does that actually mean? And what is the most effective way to check your work? Here are my top tips for checking your work efficiently and effectively:

- **Show your work.** The first step to checking your work is to have work to check. Write out your work even if you can do the math in your head. Write down the process or explanation that you're doing in your head. It'll sharpen your math communication skills today. Down the road, when you hit a math problem that is too long or difficult to do in your head, you'll be happy that you've learned how to show your work.
- **Keep a list of mistakes.** What mistakes are you making during homework and classwork? Don't check all your work on a quiz or test. Check for the mistakes you make the most frequently; the key to knowing what those mistakes are is to track them.
- **Check by question type.** Certain questions have more opportunities to make mistakes. Some questions are longer than others. Some have more calculations and simplifications. And some have more information to interpret. These are good questions to come back to and check.
- **Check word problems.** Check that you solved what the question is asking for. Did you answer the question? It's painful

to get through 90% of the work correctly but get the question wrong because you skipped the last step and didn't answer the question.

Remember that checking work is different from figuring out a hard problem. If you have a hard problem that you can't solve, should you solve it or check your work on other problems? That depends. You have to use your judgment here. Which will get you more points on your test– figuring out that tricky question or checking for mistakes? Which are you more likely to do – figure out the hard problem or catch a mistake?

Crush Math Experiment: Look back over your recent homework and classwork questions. Write down the top nine mistakes you made and the top nine question types you made mistakes on. When checking your work on your next test, check for these mistakes and these question types first.

twenty-five
when you're unprepared

I was invited to interview for an adjunct math professor position, and the timing wasn't ideal. I'd just given birth to my third child, was still in the hospital, and was more sleep deprived than I'd ever been in my life.

I did not have the time or energy to review the math I'd learned ten years earlier and was about to be tested on during the interview. I was wildly unprepared, but I showed up to the interview anyway.

Here's what I learned that day: Just show up and try your hardest while you're there. If you succeed, it'll be worth it. And, if you fail, you'll learn so much from the experience that it'll still be worth it.

If you're imperfectly prepared for your math test and there's nothing you can do about it, show up and give it everything you've got while you're there.

- Answer the questions you know how to answer.
- Use the Rule of 9 to get started when you're stuck.
- Use the Rule of 9 to check for your nine most common mistakes.
- Learn from the problems you missed and the experience, so you're better prepared next time.

I left the interview thinking I would not get the job, but I was thankful that I had learned what the interview process was like so I could do better next time. The unexpected thing was – I got the job and, once I started teaching, it didn't matter that I'd barely passed the interview. All that mattered was that I showed up and tried my hardest while I was there.

Crush Math Experiment: Next time you're unprepared for something – a test, an interview, a presentation – show up anyway and try your hardest while you're there. Afterward, write down nine things you learned from it.

twenty-six
your alter ego

In his book *The Alter Ego Effect*, pro-athlete coach Todd Herman talks about creating a heroic alter ego to tap into when you need to perform—a strategy used by many professional and Olympic athletes. As I read his book, I realized that I had been doing this for years—with math.

I had a fearless Statistics professor, Professor Janet Myhre, who led my undergraduate research group. This was my first experience with research, and, at first, I was so nervous! Every time we tried to do something new with our data, from regression analysis to the Bootstrap Method, I was freaking out.

What if it doesn't work? What if I did it wrong? What if... What if... Professor Myhre, on the other hand, was confident and unfazed by the idea of failure or mistakes:

- If you make a mistake, fix it.
- If it doesn't work, learn from it and try something else.
- If that doesn't work, learn from it and try something else.
- If that doesn't work, keep making mistakes, learning from them, and trying new things until you get to something that works.

When I returned years later to get my Master's in Math, I embraced that mentality whenever I started to get flustered. I tapped into her mathematical fearlessness whenever I didn't have it myself but needed it. I'd think, "Professor Myhre wouldn't be fazed by this." She'd just do it. She'd find a way. She'd just figure it out.

Crush Math Experiment: If you could show up to your math test as anybody, who would you show up as? What are nine qualities that they have? How would they react in the face of uncertainty or difficulty?

twenty-seven
reward yourself

One way to motivate yourself is to turn things that would usually distract you into rewards. Rewards work well for creating habits and meeting short-term goals. You can create rewards for daily, weekly, monthly, and even semester goals.

For example:

- If I finish assignment X, I will receive reward Y.
- If I get grade X on my exam, I will receive reward Y.
- If I get grade X at the end of the semester, I will receive reward Y.

Here are different types of rewards you can create for yourself.

Daily Rewards. What's your end-of-the-day reward for getting math (and your other homework) done? A TV show or video game? A special dessert? Time to talk with your friends? If you plan to do something you love after you finish your work, you can enjoy it guilt-free while staying on track for your goals.

Weekly Rewards. During my Master's program, I had tons of math homework due on Fridays. I also played soccer on Thursday nights with friends. If I got all my work done for the week by 7 pm on Thursday, I would let myself go play soccer. If I didn't finish my work, I had to skip.

Since team sports depend on players showing up, this provided tons of extra motivation to not let my team down.

Test Rewards. If your grade on your next math test is X, how will you reward yourself? Hang out with friends? Head out to dinner or a concert? Whatever it is, make it something you can look forward to, something worth the sacrifices you made to prepare.

Semester Rewards. If your final grade in math is X, how will you reward yourself? What will you do to celebrate? Don't decide at the end of the semester. Decide now, look forward to it throughout the semester, and use it to motivate you to study harder and meet your goals.

Crush Math Experiment: Think about your goal for math for the end of the semester. Make a list of nine milestones (daily, weekly, monthly, afters test, end of the semester) that you'll need to meet along the way. Pick a reward to go along with each!

twenty-eight
laziness vs. discipline

Being lazy isn't necessarily sitting on the couch and binge-watching TV. Laziness is:

- Not liking to do difficult things.
- Shying away from things that require effort or persistence.
- Staying in your comfort zone and doing things that you know you'll be able to succeed in.

Laziness is not who you are. It is a habit that you can change. Why is it so easy to be lazy? Because we are surrounded by temptation. We are surrounded by things that are easy, distracting, and vying for our attention, such as

- Netflix, YouTube, Hulu
- Email, Text, Videochat
- Instagram, Twitter, TikTok
- Minecraft, Call of Duty, Mario Kart

The list goes on and on. And it's important to pay attention to these temptations if they interfere with your ability to learn, like math, and grow as a person.

Likewise, let's take a moment to understand what discipline is. Discipline is the ability to resist temptation. It is:

- Choosing to do what you should do instead of what you want to do.
- Choosing a greater long-term payout over a smaller short-term payout.
- Choosing to study math and read this book because, in the long run, it will make you more successful, more innovative, wealthier, and better at problem-solving than Instagram or Call of Duty will.

Having a clear sense of the long-term consequences of your actions can help with making disciplined choices. What long-term benefit will you get from putting in the time and effort today to succeed in math?

Crush Math Experiment: Who do you want to be and what life do you want to build for yourself? Are your goals big and bold?Brainstorm nine ways success in your current math class will help you achieve your goals.

twenty-nine
reading for you

You may already have so much assigned reading that the last thing you want to do is read more books on your own, but the books you choose to read on your own are sometimes the most helpful.

I didn't understand this until college when, in the library, I stumbled across How to Win Friends and Influence People by Dale Carnegie. I was looking for another book and the title caught my attention. I was a freshman struggling to make new friends, so I started reading. Hours later, in the aisle of the library, I was still reading and I had this realization:

Sometimes, the most helpful books that teach you the stuff you need in the exact moment you're living in, are different from the books that are assigned in school.

Here are nine books that have helped me succeed in math or learn how to do other hard things, like teaching, writing books, starting a business, and speaking at conferences.

These are books that educators and entrepreneurs recommend to each other all the time. Still, I wish someone had recommended them to me back when I was a student. They would've helped me so much.

- *The Magic of Thinking Big* by David Schwartz helped me to create bigger goals for myself.
- *Mathematical Mindsets* by Jo Boaler helped me understand the importance of mindset in learning.

- *Can't Hurt Me* by David Goggins helped me with grit and mental toughness.
- *The Gifts of Imperfection* by Brene Brown helped me better deal with and think about my mistakes and failures.
- *How to Solve It* by George Polya helped me learn how to approach complex math problems.
- *The Pomodoro Technique* by Francisco Cirillo helped me to focus better when I worked.
- *The Five Second Rule* by Mel Robbins helped me handle my immediate reactions.

If the main thing holding you back is a fixed mindset, read a book about mindset. If the main thing holding you back is a lack of focus, read a book that teaches you how to focus. At the end of the day, you're in charge of your education and it's up to you to learn the things that will help you the most, even if they aren't covered in school.

Crush Math Experiment: What is something that you'd benefit from learning that isn't taught in school? Search for nine books on the topic and see if any are at your library.

thirty
the rule of nine revisited

If you don't like the results you're getting, change what you're doing to get those results.

How do you know what changes will be the most helpful?

You don't.

That's why you must experiment with different things until you find the things that help you the most. Here are some ideas:

- If you bombed your last test, brainstorm nine changes you could make to how you study for the next test.
- If you keep making mistakes on particular problems, do nine extra practice problems.
- If you make careless errors in basic math skills like multiplication, make flashcards and study them nine times a day for nine days.
- If you need to put in more time to finish your math homework, brainstorm nine earlier times during the week to study and nine people you can go to for help.

If you want different results in your math class, you must change how you study, think about, schedule, and do everything related to math. As in any area of your life, if you don't like the results, change your actions until you get different results.

Crush Math Experiment: If you take only one thing away from this book, it's the Rule of Nine. Brainstorm nine ways *you* can improve how *you* learn math: 9 concepts *you* can learn better, nine people *you* can reach out to, nine homework problems *you* need to revisit, etc. Then, try them out to see which ones help you the most.

thirty-one
the 80/20 rule revisited

80% of this book might not be relevant to you. 80% of the experiments you tried out might not be helpful. That's okay though because it's the 20% that we're looking out for.

The 80/20 Rule says that if you try many different things, 80 percent of your results will come from 20 percent of the things you do. Applying the 80/20 Rule to how you approach math, you'll find that:

- 20% of the experiments you try out will provide 80% of the changes you need.
- 20% of questions you ask yourself will provide 80% of the self-assessment and clarity that you need.
- 20% of the people you ask for help will provide 80% of the community and expertise that you need.
- 20% of the mistakes you track will fix 80% of your mistakes on tests.

The most important 20% of people, questions, problems, mistakes, and study skills will be unique to you. If you experiment with many things, you will find the ones that will help you the most.

Look back over the experiments in this book. Don't dwell on the 80% that was not relevant or helpful for you. Use the 20% that benefited you the most.

Crush Math Experiment: Look through the book and circle the experiments that were the most helpful to you. Write them down and come back to them whenever you need them.

chaos theory

Dear reader,

The very last thought I'd like to leave you with is this...

The road to success in math, as well as in life, is not linear. It is not a straight line with simple, easy-to-follow directions guiding you each step of the way. The road to success in math, as well as in life, is chaotic. Embrace the chaos. Crush the chaos when you can. Pick yourself up when it crushes you. It is a constant, ever-changing battle to be fought.

There is a branch of mathematics called Chaos Theory that studies the constant, ever-changing chaos in complex systems. It says that complex systems, much like your journey to succeed in math and life, are highly sensitive to slight changes in conditions. This means that teeny, tiny changes can result in crazy, humongous consequences. My hope is that the single teeny, tiny act of picking up this book will change the course of your life. I hope that it will give you that one missing tool or strategy that you need to crush math and the other challenges that life will throw at you.

I hope that math will, in turn, empower you and teach you to solve problems and conquer challenges. I hope that math will develop within you confidence, grit, and discipline. I hope that math will enable you to figure out how to do the things you think you cannot do, so that you can live a bigger, bolder, more inspired life. I hope you will crush math today, so Future You can crush anything tomorrow.

Stephen King once wrote, "Get a little rock and roll on the radio and go toward all the life there is with all the courage you can find and all the belief you can muster."

Life is short, and I encourage you to go that extra mile, live life to the fullest, and be the best that you can be. And the best version of yourself, no matter who you are, is someone who is capable of crushing math.

Love, Allison

about the author

Allison Dillard is an author, a former math professor and the host of the Allison Loves Math Podcast. She earned a B.A. in English and a B.A. in Mathematics from Scripps College, where she received the Reed Institute's Award for *Excellence in Problem Solving and Decision Science*. She has an M.S. in Mathematics from Claremont Graduate University, where she did research for Boeing, the National Science Foundation, and Los Alamos National Lab. As an adjunct math professor at Irvine Valley College, she loved teaching Prealgebra, College Algebra, and Statistics.

In 2017, Allison had health problems and took a year off teaching. After surgery and chemotherapy, she recovered quickly and the experience gave her the time and motivation to write her first book, Crush Math Now. This led to her writing more books and hosting the Allison Loves Math Podcast. Her book Crush Hypothesis Testing shows her passion for improving Introductory Statistics curriculum. Her workbooks, the Love Math Journal and

The Crush Math Experiments, show her dedication to helping educators empower their students to love and succeed in math.

Her love for math is legendary.

amazon.com/Allison-Dillard/e/B07PV43V59
facebook.com/AllisonLovesMath
instagram.com/allisonlovesmath
twitter.com/AllisonLuvsMath

gratitude

Thank you, Joe, for encouraging me to write and helping me every step of the way. You are the cosine squared to my sine squared.

Thank you, kids, for showing me how easy it is to approach life with an experimental mindset. I love you to the moon and back.

Thank you, Mom & Dad, for your support, encouragement, and problem-solving mindsets.

Thank you to my co-authors, Nicole Thomson, Jennifer Flenner, and Maria Ryan, for your expertise, originality, and vision.

Thank you to my teachers at Flintridge Sacred Heart Academy, Westridge School, Scripps College, Claremont McKenna College, and Claremont Graduate University who helped me on my own educational journey: Mrs. Mikrut, Mr. Nishiyama, Sister Ramona, Mr. Hallman, Prof. Towse, Prof. Chaderjain, Prof. Ou, Prof. Myhre, Prof. Angus, Prof. Nadim, Prof. Percus, and Prof. Rangel-Escareño.

Thank you to Irvine Valley College's math department, including Brent Monte, Joe Sheldon, JoAnne Noyes, Rich Zucker, Josh Danufsky, and Richard Shiring.

Thank you, to the trailblazers in math and education, who have helped me along the way, especially Brittany Rhodes (Black Girl MATHgic), Jonily Zupancic (Making Mathineers), Professor Prime (YouTube's Math Time with Professor Prime), Jim Frost (Statistics by Jim), Chris Skierski (Rethink Math Teacher), Carrie Cutler (Math Positive Mindsets), Audrey Codner-Gibson (ACG Tutoring), Melissa Dean (Dean of Math), Lynda Brennan (Math MileMarkers), and Punam Saxena (Edu-Me Podcast).

Thank you, moms, especially Megan Haskell, Catherine Currie, Beth Marcus, Alicia Hice, Vinky Winkler, Erika Chico, Amy Vallejos, Tracy Campbell, Sydney Oster, Seema Ahmad, Casey Adams, Julia Miles, and Viviana MacManus, for your support.

Thank you to the Mathematical Mindset series authors, Jo Boaler, Cathy Williams, and Jen Munson for laying the foundations for a generation of educators to learn about and teach mindset in mathematics.

raise your math grade

You can get a free copy of Allison's book, Raise Your Math Grade, at allisonlovesmath.com.

also by allison dillard

The Math Anxiety Handbook (2023)

Abby Solves the Hungry Kitten Problem

Crush Hypothesis Testing

The Love Math Journal

Raise Your Math Grade

Crush Math Now

Available for Purchase on Amazon, Barnes & Noble, and other online retailers.